Electronics are only for nerds and couch potatoes?

Not at all: whether with LEDs, motors or solar cells, this book shows how exciting electronics can be!

Want to build a spaceship with LED lighting? A solar-powered boat? A robot mask? Then you've come to the right place.

We have a number of projects for indoor and outdoor experimenting, which can be completed in just a few steps. All electronic parts required are included with this book; the projects combine these parts with everyday household items that you are certain to find at home.

These projects run on a very safe and low electrical voltage of 3 volts, or use a solar cell for power. However, make sure you get an adult to check that you have followed all instructions correctly and to help with any fiddly project bits.

Ready to dive into Electronic Adventures?
Then let's go!

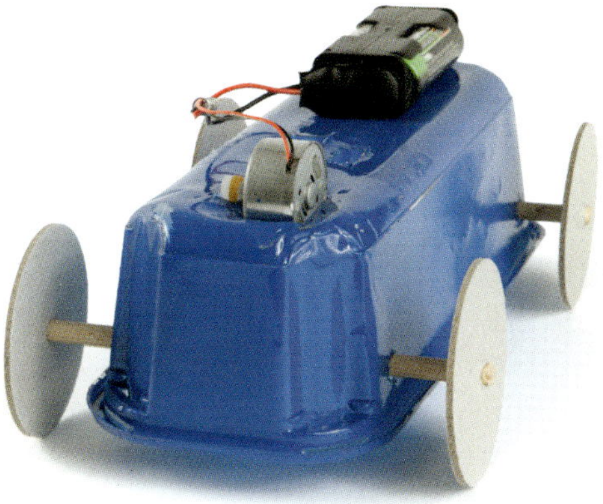

3

Parts

Your electronic kit

Crocodile cable

Cables with crocodile clips are extremely useful; you can attach them to the screw of a bulb socket and to stripped cable ends. Crocodile cables are used instead of switches in many projects.

Cable

The current has to flow through something; you just cut the correct length for your project, remove the insulation at the ends (see page 31) and twist the wires around the connections on your project part.

The cable colours have specific meanings: **red** stands for positive, **blue** or **black** for negative.

Bulbs

Essential for any project where you want to light up something.

Bulb socket

You need the bulb sockets for the coloured bulbs! Screw the bulbs into the sockets and screw stripped cable ends onto the screw connections or simply connect a crocodile cable.

Rainbow LED

Rainbow LEDs are great light-emitting diodes. Once connected, they change colours from red to green to blue. But beware: in contrast to bulbs, when you connect an LED you have to make sure that the current is flowing in the right direction. The longer pin (anode) must be connected to the positive pole, and the shorter pin (cathode) must be connected to the negative pole.

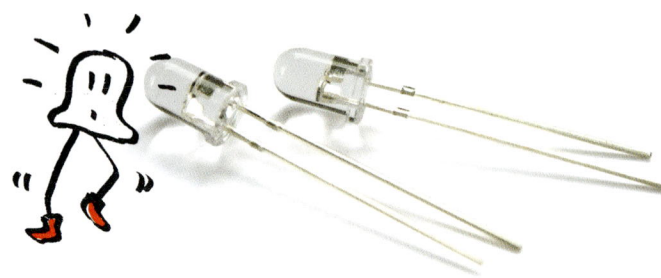

Switches

A necessary part for projects that have to be switched off from time to time. Professionals solder the cable ends, but we think it works just as well when you wrap a stripped cable end around the connections.

DC motor with fitting

Essential for anything that needs to move. Your motor has two cables, one red and one black. The red cable is connected to the positive pole of your power source, and the black one to the negative pole. Anything you want to turn into a moving part can be attached to the motor's spinning axle and motorised. You can push the fitting onto the axle to widen it.

Solar cell

A power source for your green projects; your solar cell has a red (positive pole) cable and black (negative pole) cable, which are connected to the motor connections.

Fan

You can use this to generate wind power. Just stick the fan onto the motor axle and use it to power vehicles or boats.

Battery holder

This is your most important power source. Insert two 1.5 volt batteries (LR 6/AA/mignon cell) and you'll have 3 volts of electricity available at the wires attached.

Cardboard disc

Provides a central and stable platform for attaching other objects to the motor in many projects.

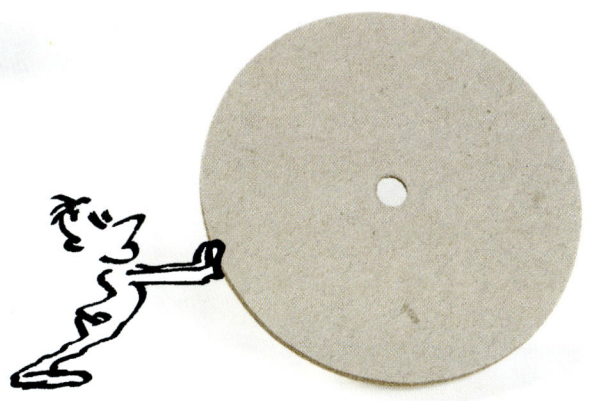

Contents

FRANZIS

Safety information for parents and kids

Warning! Not suitable for children under 3 years of age. There is a choking hazard, as small parts may be swallowed or inhaled.

Warning! Only suitable for children at least 8 years of age. Instructions for parents or other responsible adults are included and must be followed. Retain the packaging and instructions as these contain important information.

Warning! Do not conduct any experiments with electrical outlets. The 230 volt mains presents a risk of electrocution.

The battery holder requires two 1.5 volt batteries (LR 6/AA/mignon cell), which are not included with this book due to their limited storage conditions.

Avoid a short circuit of the batteries as this may cause the cables to overheat and the batteries to explode.

Mixed battery types, such as rechargeable and non-rechargeable units or new and used batteries, must not be used together in the battery holder.

Batteries must be used with the right polarity and pushed into the holder without too much force.

No attempt should be made to charge non-rechargeable batteries as this leads to a risk of explosion.

Chargeable batteries should only be charged under supervision of an adult.

Do not use rechargeable batteries as these may explode in the case of a short circuit.

Remove empty batteries from the battery holder.

Used batteries must be disposed of in accordance with legal provisions.

Batteries must not be connected to metal objects, thus causing a short circuit.

Avoid battery deformation.

All of the projects in this book use the battery holder supplied (requiring two 1.5 volt batteries, LR 6/AA/mignon cell) or the solar cell. These projects therefore run from a very safe and low electrical voltage of 3 or 4.5 volts.

Explicitly instruct your child to read and carefully follow all instructions and safety information. Project instructions and rules must be followed.

Have fun building
and experimenting!

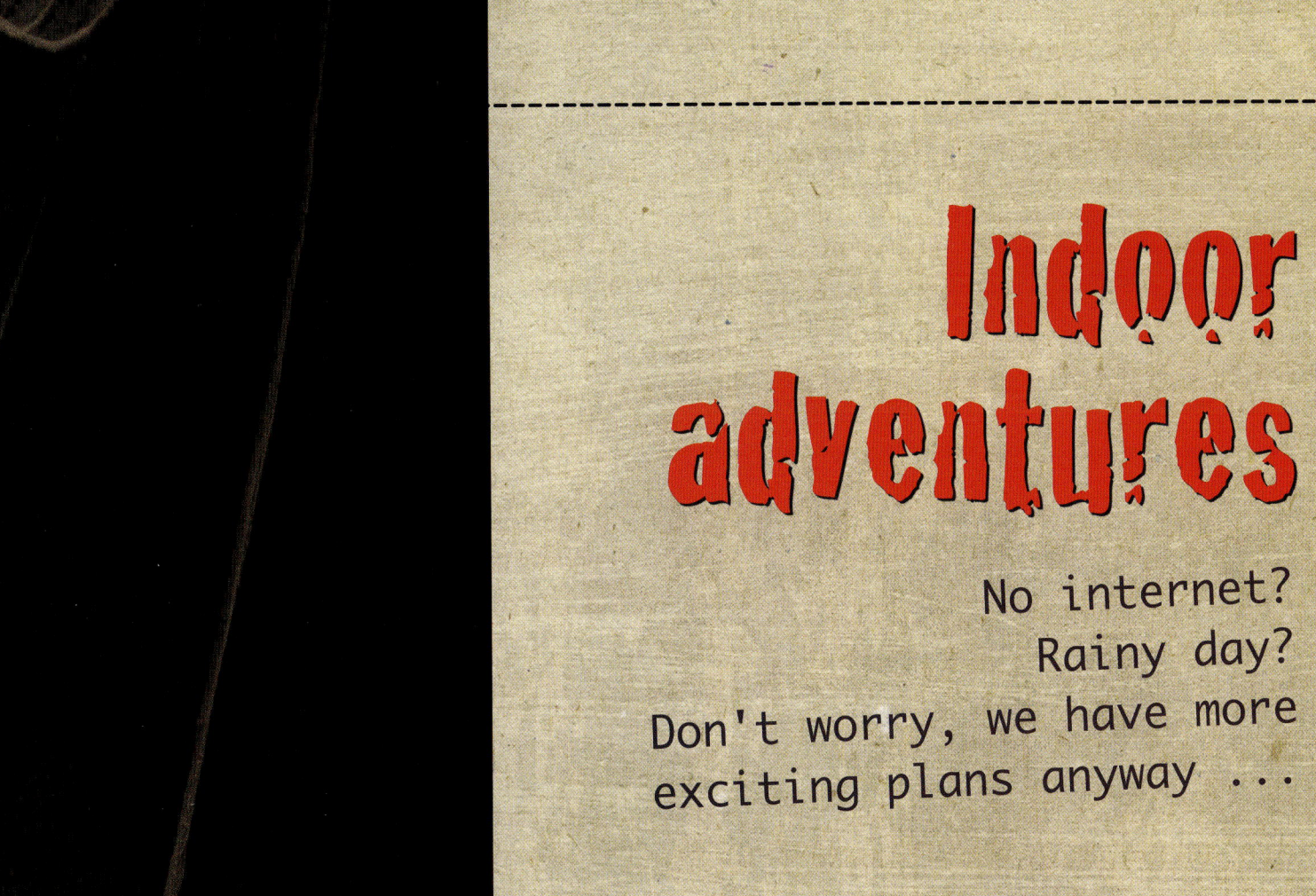

Indoor adventures

No internet?
Rainy day?
Don't worry, we have more
exciting plans anyway ...

Treasure ahoy!

If Blackbeard had only known ...

Fill up your treasure chest, mateys!
Use electricity and salt water to make your pennies look like they have been buried in the ground for hundreds of years.

You will need:

2 crocodile cables

Plus:

1 4.5 V battery
Note: You can use the 3 volt battery holder instead. But it's faster with a 4.5 volt battery.

1 small, shallow glass dish

1 long screw

salt

warm water

1 handful of copper coins (1p or 2p coins)

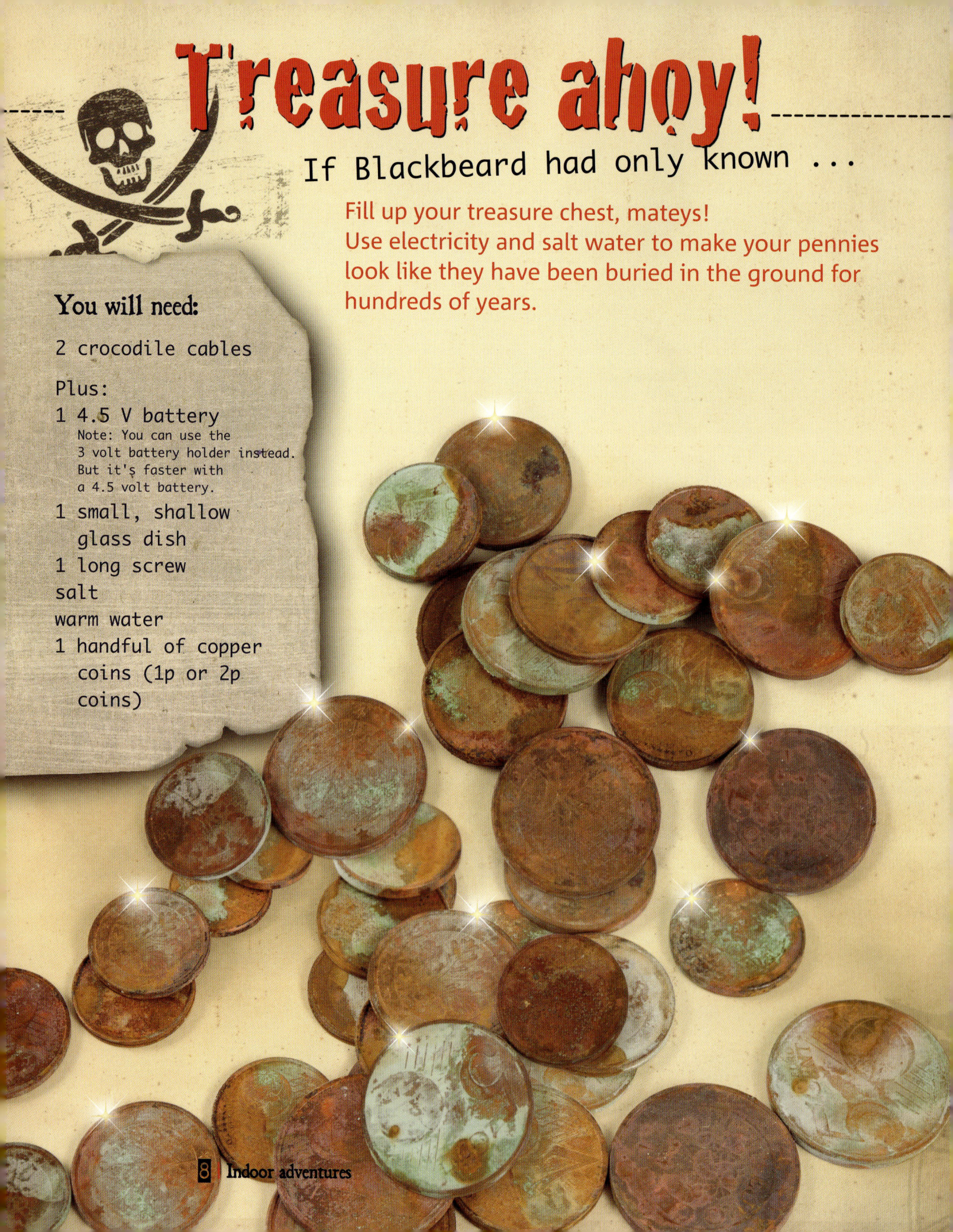

1. Fill a shallow glass dish to the top with warm water and add one teaspoon of salt.

2. Attach one crocodile cable to the screw and hang it over the edge of the glass dish. Attach the other crocodile cable to a penny piece and hang the coin over the opposite edge of the glass dish. The coin should be covered by the water as much as possible.

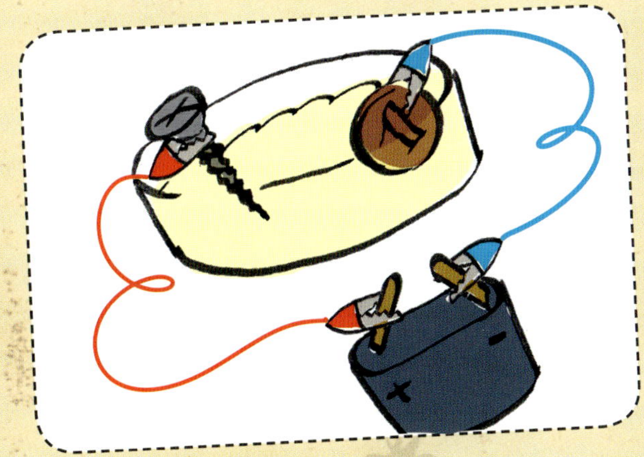

3. Connect both crocodile cables to the battery as shown in the picture.

Now you just have to wait. After a short time, bubbles will form on the screw and the water will turn a dark yellow-green colour. After a few minutes, the penny will become dull and green. If you wait even longer, the engravings on the coin's surfaces will be 'eaten away' completely.

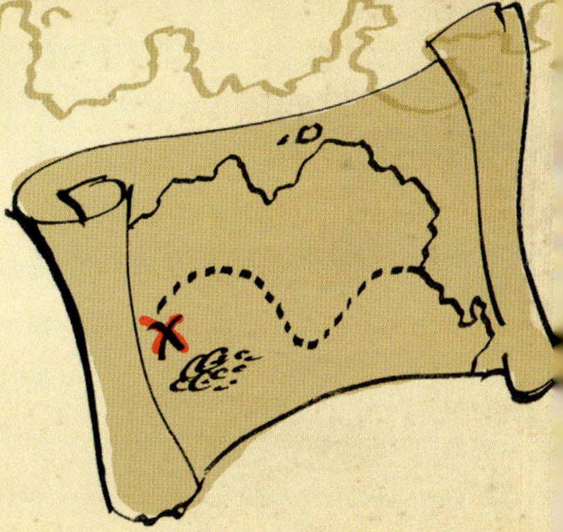

After treating two coins, you should throw out the dark water and replace it with a new salt water solution.

We have lift-off!

Explore space at light speed

Just about anyone can build a spaceship. But if you want to explore space with cool LED lighting, it takes special planning.

You will need:

1 switch
1 battery holder with two 1.5 volt batteries
approx. 50 cm of red cable and 50 cm of black cable
2 rainbow LEDs

Plus:
sturdy cardboard
empty tubes from kitchen and loo roll
2 pairs of lustre terminals
a thick needle
paint for colouring your ship

1. Make a space ship out of cardboard and empty tubes from kitchen and loo roll. Use adhesive tape and a hot-glue gun to attach the spaceship parts.

5. Now connect the LEDs, lustre terminals, switch and battery pack together as shown in the picture. Make sure that your cable is long enough to be directed through the spaceship. If you need a longer cable, go to page 31 to find out how to extend your cable.

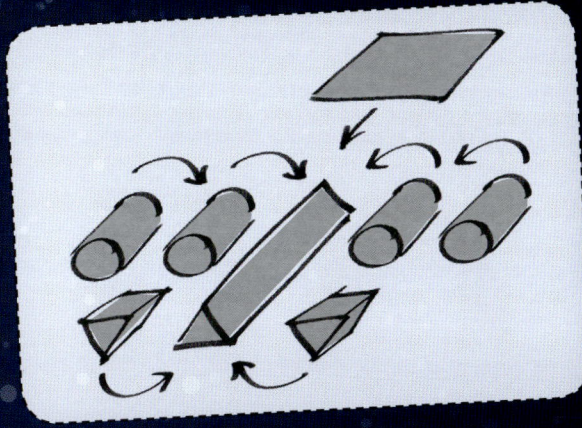

2. Attach a platform with seats for the ship's crew and paint the spaceship any colours you want.

3. Choose two easily accessible spots for your LEDs. Use a thick needle to create two holes for the LED pins, leaving a distance of 3 mm between them. Do this for each LED. Now insert the LED pins into the openings and mark the holes with the longer pins as the positive poles.

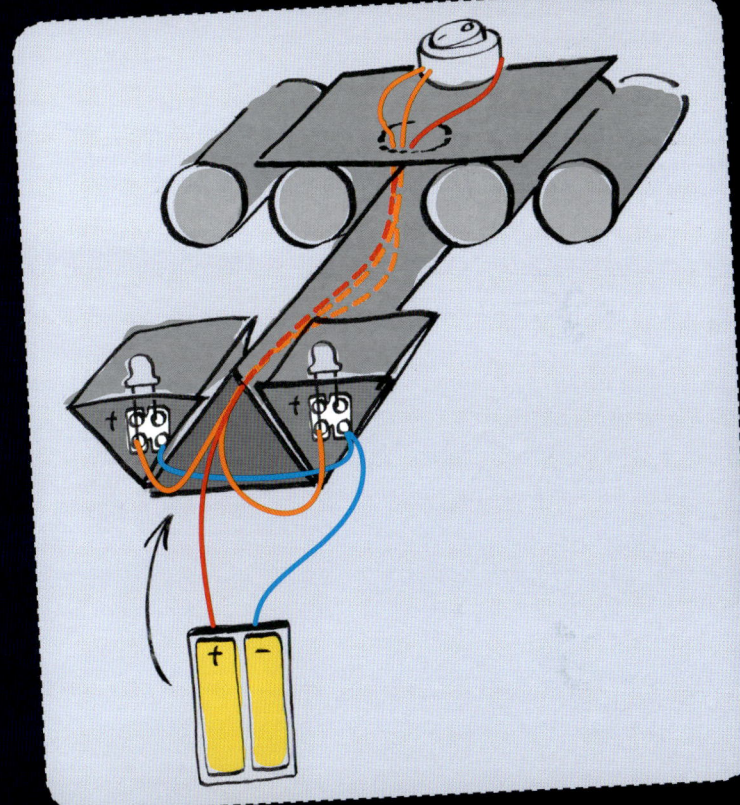

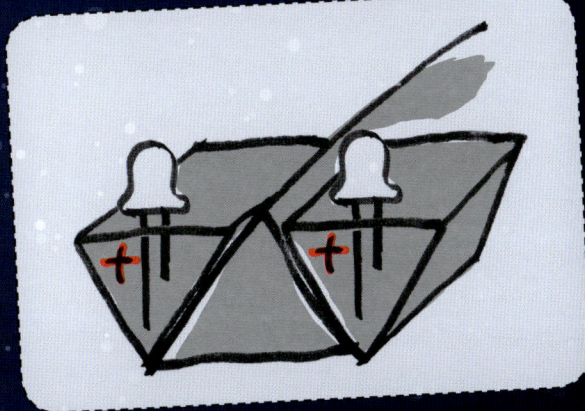

6. Connect the cables to the switch and then push the switch into the opening from the outside.

7. Now hide the battery pack inside the ship. Ready for lift-off? Then let the space exploration begin!

Tip: These lights look good on other flying objects, too, as well as model cars and pirate ships.

4. Choose an easily accessible place for the switch and cut a 2 cm wide opening.

Keep out!

No adults allowed!

There are times when you just don't want mum, dad or your little brothers and sisters in your room. Make a door sign with lights to signal to your visitors whether they are welcome or not.

You will need:

2 bulb sockets
1 red and 1 green bulb
1 battery holder with two 1.5 volt batteries
door sign template (see page 58)
4 cables, each one 7 cm long
2 crocodile cables

Plus:
1 piece of sturdy cardboard, size A4
screwdriver
about 5 cm of string
adhesive tape

VISITORS WELCOME

KEEP OUT!!!

1. Copy the door sign template on page 58 onto A4 paper and glue it to the cardboard.

2. Use sharp scissors to cut out the marked circles to create holes for the bulb sockets.

3. Cut four bits of cable to about 7 cm in length and remove 1 cm of the insulation at the ends (see page 31). Create a loop on one end of the stripped cable and carefully screw it into the screws on the bulb sockets, so that each socket has two cables hanging from it.

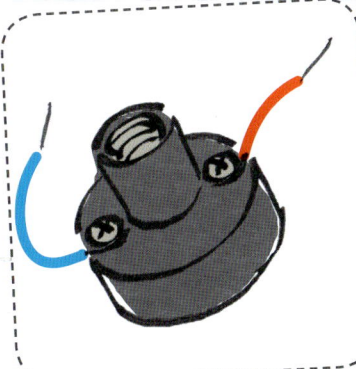

4. Insert the bulb sockets through the holes from the back of the sign. Screw the red bulb into the socket next to the words 'KEEP OUT!!!' and the green bulb into the socket next to the words 'Visitors welcome'.

5. Use the piece of string to make a hanger for your sign and attach it to the back of the sign using adhesive tape.

6. Use adhesive tape to stick the battery holder with the crocodile cables and batteries to the back of the sign.

7. No entry allowed? To switch on a light, use a crocodile cable to connect the red wire from the battery holder to the red wire for the socket with the bulb you want to turn on. Use the second crocodile cable to connect the black wire from the battery holder to the black wire for the bulb socket. If you want to switch the sign completely off, just disconnect one of the crocodile cables and let it dangle freely.

Warning! The connection contacts for the red and black cables should not touch, otherwise a short circuit will occur. Carefully fix cables into place using adhesive tape.

The robots are coming!

Is it Halloween again already?

No! But you can wear robot masks any time of the year. Only here will you find one with eyes that light up in the dark.

You will need:

1 battery holder with two 1.5 volt batteries
2 sockets with bulbs
4 crocodile cables
4 cables, each one 20 cm long

Plus:
1 cardboard box that fits over your head, at least 35 cm high
spray paint
2 small yoghurt pots
coloured plastic film
black mesh material
scissors

1. This mask uses the robot's mouth as the eye slot. First turn the box over and mark where your own eyes will go. In this spot, draw a long rectangular opening for the mouth. Then use the yoghurt pots to draw the circles above the mouth for the robot eyes. Use sharp scissors to cut the openings. The eye holes must be cut a little smaller than the yoghurt pots so that they will not fall into the box. Create four small holes at the back of the box, opposite the eye holes.

2. Now spray the cardboard box with silver spray paint. Get an adult to help you with this, and make sure you never breathe in the spray paint.

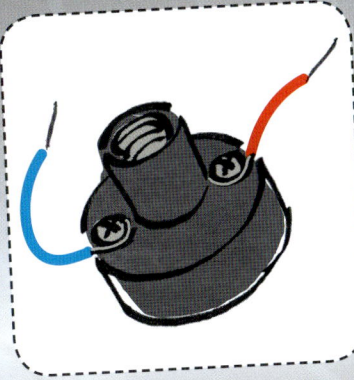

3. Cut four bits of cable to about 20 cm in length (two red and two black) and remove 1 cm of the insulation at the ends (see page 31). Connect the stripped cable ends to the bulb sockets.

4. Use sharp scissors to make two holes in the bottom of both yoghurt pots. Screw the bulbs into the sockets. Place the sockets into the yoghurt pots and direct the cables through the holes in the bottom of the pots and then out through the back of the cardboard box. Stick them to the box with adhesive tape.

5. Stick the coloured plastic film circles over the eye openings. Inside ... attach a strip of black mesh material over the robot's mouth opening.

6. Use adhesive tape to stick the battery pack onto the back of the cardboard box. To switch on your robot mask, use the crocodile cables to connect the battery pack and bulb cables as shown in the picture.

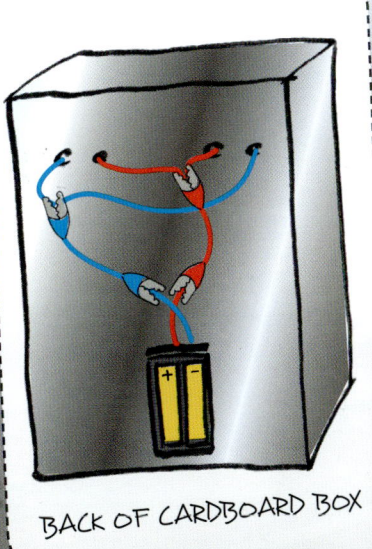

BACK OF CARDBOARD BOX

Warning! The connection contacts for the red and black cables should not touch, otherwise a short circuit will occur. Carefully fix cables into place using adhesive tape.

Action painting

Warning: this one's messy!

With rotating discs, you can use paint and brushes to create the most amazing designs. But be careful! The paint may splash and cause a big mess.

You will need:

1 motor with fitting
2 crocodile cables
1 battery holder
    with two 1.5 volt
    batteries

Plus:
white card
compass for making
    circles
scissors
water colours and
    paintbrush or
    felt-tip pens
strip of corrugated
    cardboard as a spray
    guard
thin string
tree branch
leather strap

Pendants

If you use very thick card or even wood to create your discs, you could turn your painted creations into pendants. Create a hole at the edge of the disc, attach a metal link (from an old key ring, for example) and thread it onto a leather strap.

1. Use the compass to draw circles on the card, cut out the circles and create a hole in the middle that is just under 2 mm.

2. Put a card circle (disc) on the motor axle with the fitting in place and connect the motor to the battery pack. The disc should rotate as smoothly as possible, so that the paint can be spread evenly.

SPLASH!

3. Shape the corrugated cardboard into a spray guard, as otherwise the paint could splash onto your clothing or the wall. You could go without the spray guard if you wanted to redecorate your room this way, but always get permission from your parents first.

4. Make coloured rings on the rotating disc using the water colours, by dragging the brush along the disc as it turns.

Alternatively you could use thick felt-tip pens to create your own original designs.

Peace maker!

Make a peace mobile

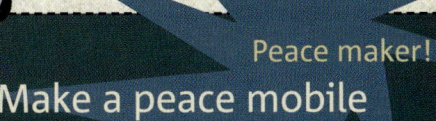

If the paint accidentally gets onto your clothing or the walls and your parents are not happy about that, turn your colourful discs into a beautiful mobile to restore peace. Create small holes at the top and bottom edge of the discs, connect them with string and hang them from a bit of tree branch.

Racers, boats & carousels

Full steam ahead!
We won't get
out of the water 'til the
sun goes down!

Solar pirates

Build your own boat!

No wind for days? No problem!
Sail away with your solar-powered catamaran.

You will need:

1 solar cell
1 motor

Plus:
2 empty, plastic
  500 ml bottles with
  caps
1 wooden board
  about 17 x 8 cm
4 cable ties
card
2 drawing pins
adhesive tape
stapler
electrical tape

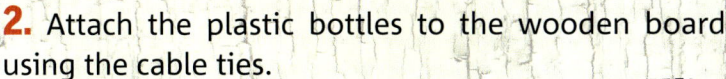

1. Get an adult to drill four holes with a diameter of about 6 mm into the corners of the wooden board. (The cable ties must be able to fit through the holes.)

2. Attach the plastic bottles to the wooden board using the cable ties.

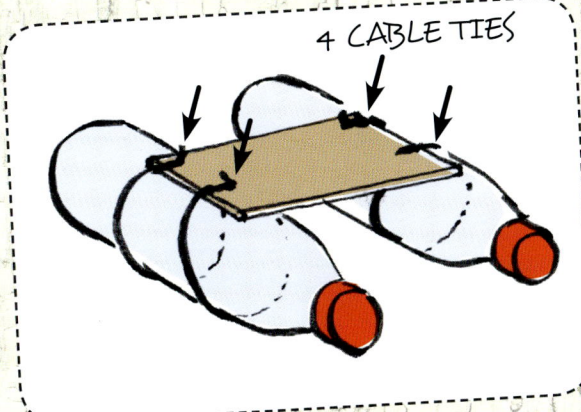

4 CABLE TIES

3. Now take a medium-strength card and cut a strip 2 cm wide and about 18 cm long. Wrap the middle around the motor and staple the strip together, so that you end up with a loop of card about 8 cm wide, through which the motor and fan can fit.

STAPLE!

4. Use two drawing pins to stick the motor holder onto the back part of the wooden board.

5. Use two bits of adhesive tape to stick the solar cell onto the wooden board. Make sure that the adhesive tape only covers the edge of the solar cell.

6. Twist the red motor cable around the solar cell's positive pole cable and the black cable around the negative pole cable. Wrap the twisted ends with a piece of electrical tape.

ELECTRICAL TAPE

Now you can turn your catamaran into a pirate ship. Of course, you'll need a pirate flag! You can find a pirate flag template on page 59.

Is the sun shining? Then your solar boat is ready for its first voyage!

If your fan starts turning in the wrong direction, just swap the cables on the solar cell.

Super speed racer

Who can make
the fastest electric car?

You will need:

1 motor with fitting
1 battery holder
   with two 1.5 volt
   batteries

Plus:
4 card wheels with
   5 mm holes
about 20 cm wooden
   sticks with a
   diameter of 5 mm
6 balloons
1 small rubber band,
   about 2.5 cm in
   diameter
1 empty plastic
   container, such as
   a tomato punnet from
   the supermarket
1 spring from a
   ballpoint pen
scissors
saw
adhesive tape

1. Use sharp scissors to cut four openings on the sides of the plastic container. The openings must be large enough for the wooden stick to fit into and spin easily.

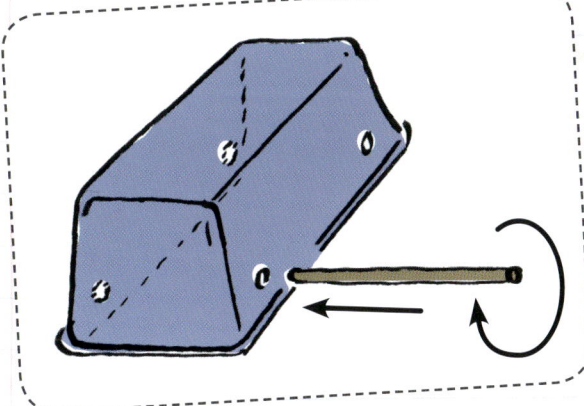

2. Use the saw to cut two axles of equal length from the wooden stick. They should stick out from the plastic container about 1 cm on both sides.

3. Use the diagram below as a guideline for cutting the openings in the container for the motor and the fitting. The position of the openings depends on the dimensions of the plastic container; if the container is very high, the motor opening can be placed directly over the rear axle. If the container is more shallow, the opening should be positioned in the middle of the container.

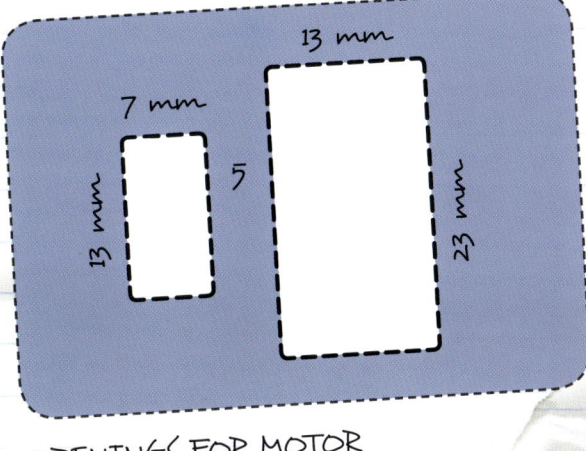

OPENINGS FOR MOTOR

4. Take the balloons and cut the thick ring at the opening off. Twist two balloon rings twice around the front axle and four rings twice around the rear axle.

5. Loop the rubber band onto the rear axle and insert both axles into the plastic car body. Stick the wheels into the axles and adjust the balloon rings so that they are in the same positions as in the picture. The rubber band is used to connect the motor and the axle and should run between the middle balloon rings in a straight line upwards to the fitting on the motor.

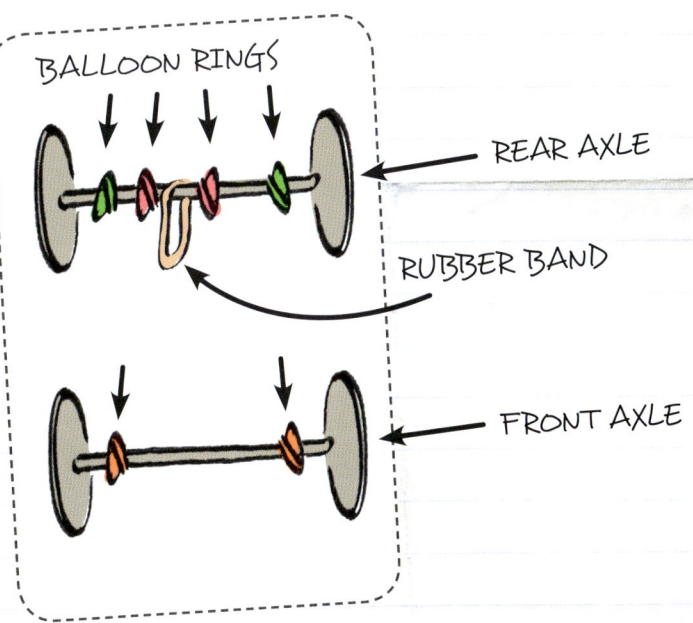

BALLOON RINGS

REAR AXLE

RUBBER BAND

FRONT AXLE

6. Make sure that the fitting is inserted on the motor upside down, with the wider end facing away from the motor. Push the motor into the opening on the container from the top. Loop the rubber band, which is connected to the rear axle, onto the fitting. Make sure that the rubber band is taut and that the motor axle can spin freely. You may have to use a smaller rubber band or cut a bigger opening for the fitting.

RUBBER BAND

7. Use a bit of adhesive tape to stick the battery holder (with batteries inserted) onto the car body. Twist the two black cable ends together. To switch your car on, stick the two red cable ends into the spring from a used ballpoint pen.

VROOM, VROOM!

Tuning tips

Having problems transferring power from the motor to the wheel axle? Make sure the rubber band is taut enough; use a smaller rubber band if necessary. Make sure that the motor axle can spin freely through the opening on the container and is not being blocked by something. Your car will run best on a smooth surface, such as tiles or wooden floors.

Stay cool

Need to cool down?

If the summer heat gets to be too much, make a Native American-inspired solar-powered fan to help cool down.

1. Cover the empty salt box with coloured paper or paint it whatever colour you want.

2. In the top third part, cut a round opening as a motor holder. Make the opening a little too small and make little cuts around the edge so that the motor will only fit if it is pushed in very tightly.

3. Fill a third of the box with pebbles so that it will be stable when placed upright.

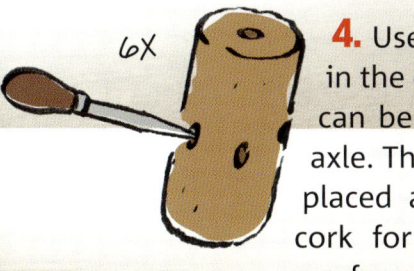

4. Use the awl to make a hole in the end of the cork, so that it can be pushed onto the motor axle. Then make six holes evenly placed around the sides of the cork for the feathers to create your fan.

5. Push the cork onto the motor axle. Then push the motor through the opening in the box from the outside, applying pressure from the inside at the same time so that it does not fall into the box. Direct the cables from the motor through the top opening of the box and tape the lid closed with a bit of adhesive tape, allowing the cables to remain outside of the box.

6. Twist the red cable from the motor around the red cable for the solar cell and wrap the connection point with some electrical tape. Do the same with the two black cables. Stick the solar cell to the 'roof' of the box using a bit of adhesive tape. Make sure that the adhesive tape only covers the edge of the solar cell. When light hits the solar cell, the fan will begin to turn.

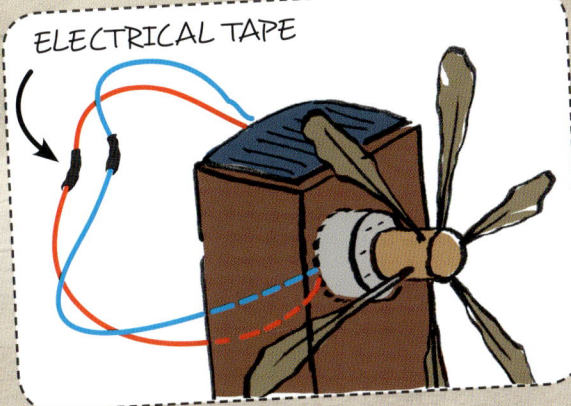
ELECTRICAL TAPE

Yoghurt pot model

This fan is made in the same way as the Native American version, except that with this model, you first eat a yoghurt and then clean the empty pot. Then cut the yoghurt pot into six evenly sized parts and separate them from the bottom of the pot. Cut six slots 5 mm deep into the cork with a knife, stick the yoghurt pot parts into the slots and then secure them with glue.

Real action for your figures

Give your
action figures a ride on a carousel!

Use an empty drink can
and a coaster to make a solar-powered carousel.
Place on a windowsill for best results.

You will need:

1 solar cell
1 motor with fitting
1 cardboard disc
2 crocodile cables

Plus:
1 empty energy drink
  can (these are the
  slim cans)
1 mug of sand
double-sided and
  normal adhesive tape
1 drink coaster
coloured paper
  or paints
thin string
4 beads
4 large paper clips
some modelling clay
4 action figures to ride
  your carousel
scissors
glue

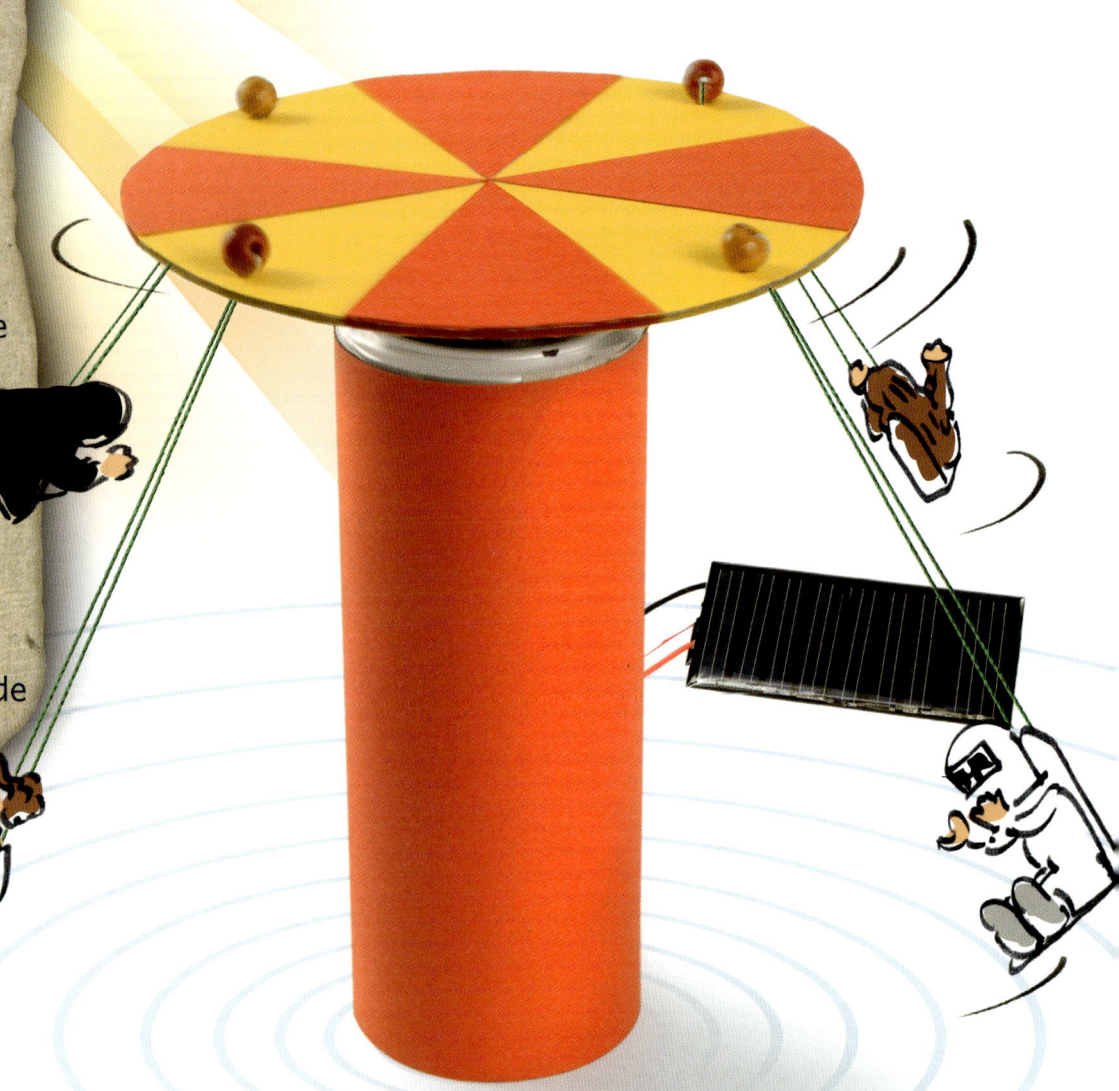

1. You have plenty of energy already, so get a tired adult to drink the energy drink and return the empty can to you. Remove the tab from the can, fill the can with sand and then close it up with a bit of modelling clay. Then cover the can with coloured paper or paint it.

2. Now turn the can upside down and fill the curved bottom of the can with modelling clay.

MODELLING CLAY

3. Colour or paint the drink coaster however you want. You could glue different stripes or segments of coloured paper onto the coaster.

4. Mark the centre of the coaster. See page 31 to find out how to find the exact centre. Use scissors to make four evenly placed holes with a diameter of 2 to 3 mm around the coaster. These holes will be used to hang the carousel seats.

5. Now use double-sided adhesive tape to stick the cardboard disc to the underside of the coaster, right in the middle, and then poke an opening in the coaster through the hole in the middle of the disc.

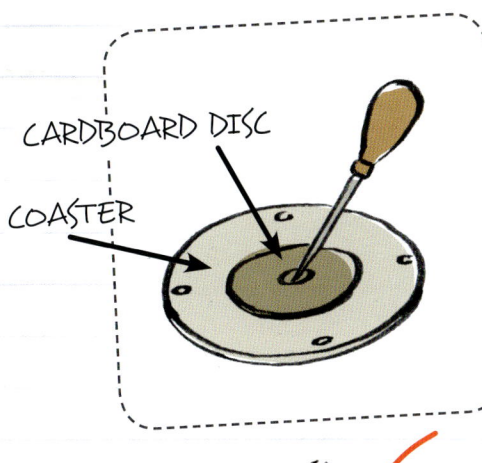

CARDBOARD DISC

COASTER

6. Place the fitting on the axle of the motor and press the motor into the modelling clay.

7. Carefully push the carousel disc onto the motor axle. Tape the cables from the motor flat against the side of the carousel stand, so that they don't hit the seats later. Make sure the cables are about 1 cm apart from each other right at the bottom so that they cannot come into contact.

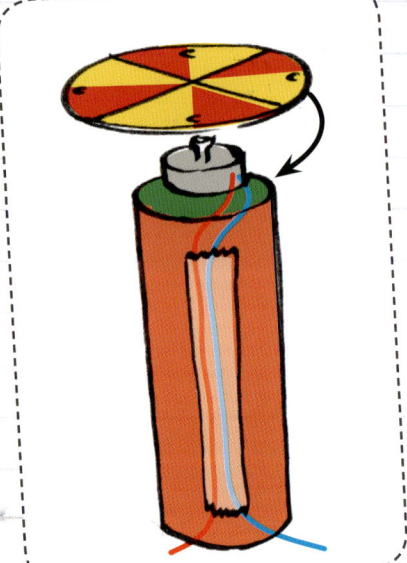

8. Shape the paper clips into seats by bending the inner part out. Use adhesive tape to stick your action figures to the seats.

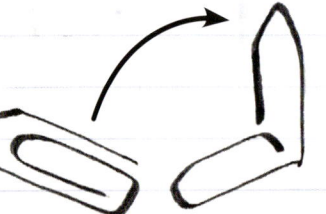

9. Now tie a piece of string to a seat. Push the free end of the string through an opening on the carousel disc, thread a bead onto the string and then feed the string back through the opening and tie the end securely to the seat. Get an adult to help with this, because it is not very easy to make sure that all of the seats are hanging at the same height.

10. Now the only thing missing is the power source. Use the two crocodile cables to connect the solar cell to the motor as shown in the picture. When enough light hits the solar cell, your carousel will begin to turn.

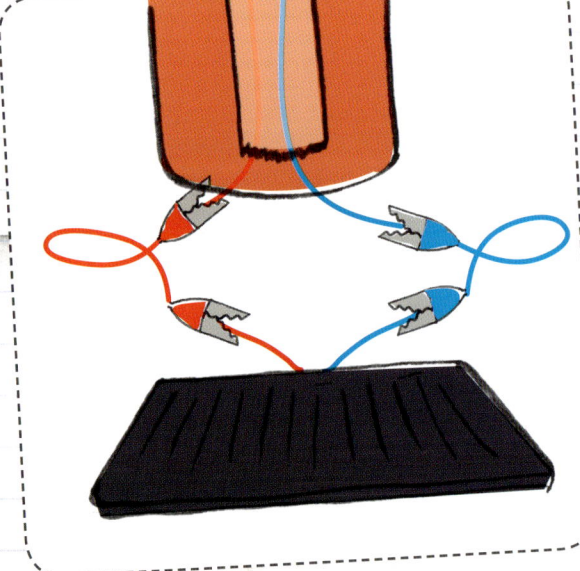

If your carousel starts turning in the wrong direction, swap the cables on the solar cell before your action figures get sick!

Stripping cables

To connect a cable, you must first remove about 1 cm of the insulation. The insulation is the coloured plastic covering around the copper wire. Professionals use insulation-stripping pliers, but craft scissors work well, too. Carefully press the scissors around the insulation, so that you cut it without cutting the wire inside. Then remove the insulation using the scissors or your fingernails. If you press too firmly, you will cut the wire itself, so first practise a few times or get an adult to help you if necessary.

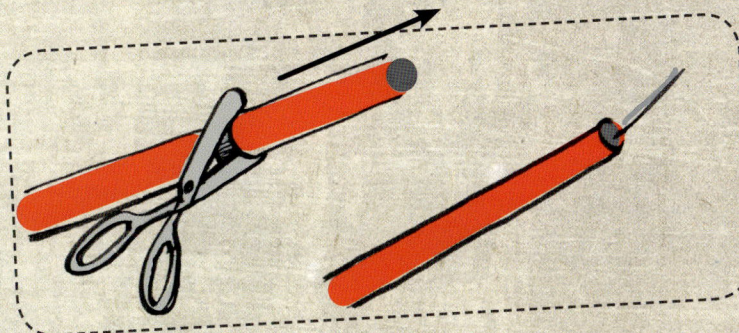

Screwing in cables

The cables are screwed into the screw connections on the bulb sockets. Create a loop at the stripped end of the cable, place it around the screw on the socket and then tighten the screw with a screwdriver.

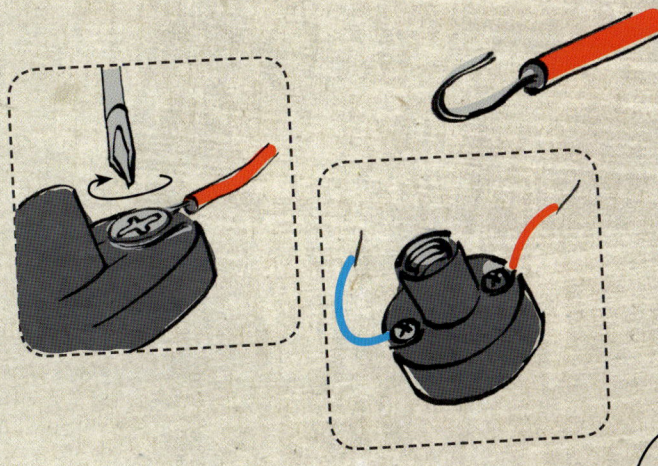

Attaching cables to the switch

Your switch has two connections. To connect the cables, direct the stripped cable end through the hole in the terminal lug and wrap it around the metal indent.

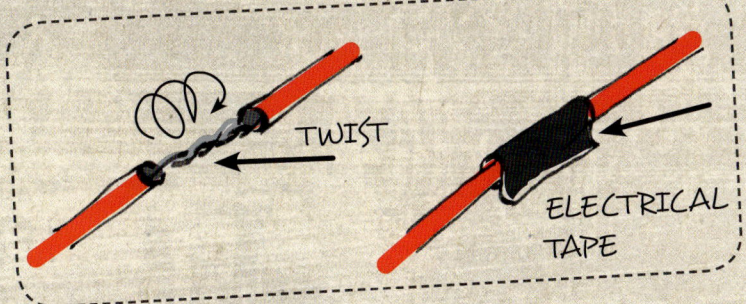

Cable too short?

If you need a longer cable for your project, you can connect pieces of cable together. Twist the two stripped cable ends together and wrap a bit of electrical tape around them, if you have any.

TWIST

ELECTRICAL TAPE

Finding the centre of a circle

How do you find the exact centre of a round drink coaster or cheese box? Lay the coaster on a white piece of paper and trace around it with a pencil. Cut out the circle from the paper and fold it exactly in half twice, then unfold again. The point where the two fold lines meet is the centre. The paper circle can now be placed on the coaster and used as a template to mark the centre of the coaster.

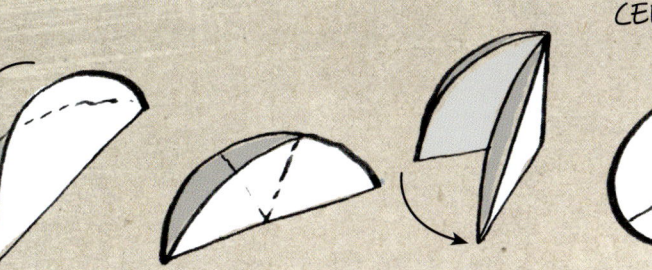

CENTRE

Sleepover party

We're not tired!
Even after the sun goes down,
the ideas keep on coming.
In fact, lots of things
are more fun in the dark.

Not just for coppers

When is the next sleepover?

You probably have a torch somewhere around the house. But you never seem to find one right when you desperately need one. So why don't you just make your own!

You will need:

1 battery holder with two 1.5 volt batteries
1 bulb with socket
1 switch
2 cables, one red and one black, 30 cm long each

Plus:
1 large (!) round sweets tube – these are the perfect size for the socket
black adhesive film
1 lustre terminal
scissors
screwdriver

1. First eat all of the sweets so you have an empty tube.

2. Cover the tube with black adhesive film and cut a 2 cm opening near the tube cover for the switch. Use sharp scissors to make two holes opposite each other in the bottom of the tube.

3. Cut two lengths of cable, one red and one black, about 30 cm each, and remove the insulation at the ends (see page 31). Now screw both cables to the screws on the bulb sockets.

4. Push the cable ends through the holes from the outside of the tube and direct the cables to the other end of the tube. Push the bulb socket carefully into the deep part of the bottom of the tube.

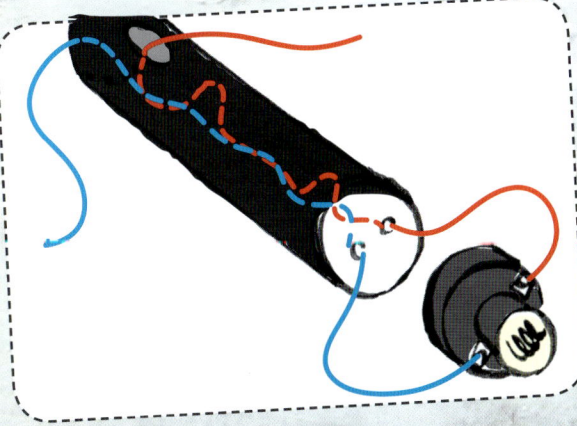

5. Connect the two red cables to the switch as shown in the picture. Connect the two black cables using a lustre terminal.

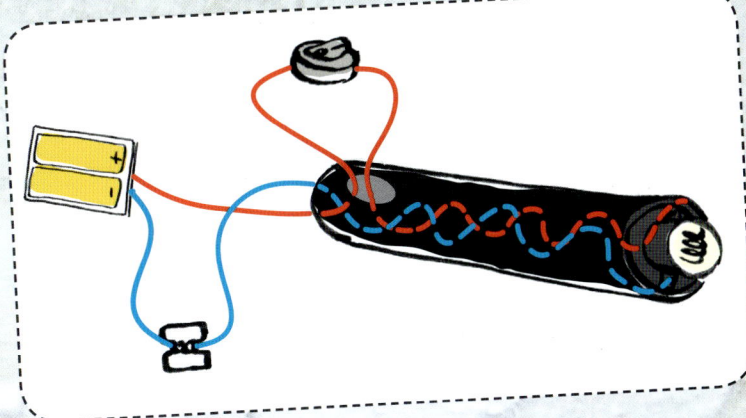

6. Press the connected switch into the opening from the outside. Insert the batteries into the holder and carefully push the entire battery pack into the torch. Put the cover on and switch on your torch!

Tip: To get more light, you can make a funnel-shaped reflector and attach it to your torch. Cover the inside of the funnel with aluminium foil. Make sure the foil does not touch the bulb socket connections.

The witching hour at the secret meeting place

Trick or treat

Scare them all off!

A scary pumpkin that changes colours is guaranteed to scare off any ghost hanging about – and maybe even your grumpy neighbours too!

You will need:

1 rainbow LED
1 battery holder
   with two 1.5 volt
   batteries
2 crocodile cables

Plus:
1 small pumpkin
knife
spoon
felt-tip pen
a polystyrene
   block

4. Spread the prongs of the LED and stick it into a polystyrene block. Mark the point where the longer prong is with a '+' for the positive pole. Connect a crocodile cable to the positive pole of the battery pack and connect it to the positive prong on the LED. Use the other crocodile cable to connect the negative pole of the battery and the shorter prong of the LED.

1. Cut the top of the pumpkin off. This part will be used as a cover later so don't get rid of it.

2. Scrape out the insides of the pumpkin with a spoon, leaving a 2 cm thick wall.

3. Use a felt-tip pen to draw a scary face on the pumpkin and carefully cut out the face with a knife.

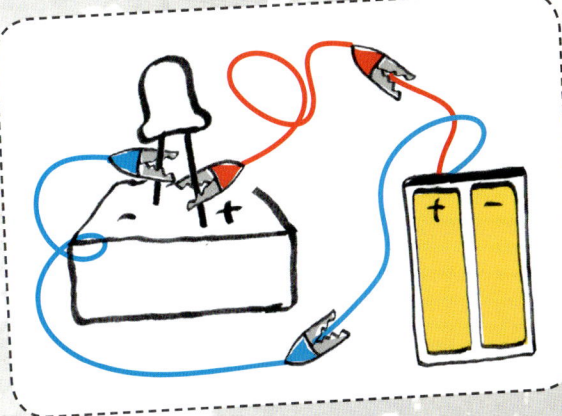

5. Now place the polystyrene block and battery pack into the pumpkin and place the cover on top. To switch off the pumpkin, disconnect one of the cables.

Warning! The connection contacts for the red and black cables should not touch, otherwise a short circuit will occur. Carefully fix cables into place using adhesive tape.

Devilish deeds

Simply terrifying!

Banish nightmares and your younger brothers and sisters all at once! A twinkling mask with eyes that change colour from green to blue to red is a real eye-catcher for your room.

You will need:

1 switch
1 battery holder
  with two 1.5 volt
  batteries
12 cm red cable
6 cm black cable
2 rainbow LEDs

Plus:
1 empty cheese box
  with a diameter
  of about 12 cm
coloured paper,
  scissors, glue,
  screwdriver,
  1 needle
2 pairs of lustre
  terminals

1. Glue a face made out of coloured paper onto the cheese box. It could be a devil face, or even a tiger or a clown, if you want something less scary. Make two holes in the upper edge of the box and attach a bit of string to hang your mask.

2. Use the switch as a nose for your mask. Mark the point and cut a 2 cm opening. Mark two points on each eye for the LEDs, leaving a distance of 3 mm between the points. Use a thick needle to make holes for the LED prongs.

3. Now carefully stick the rainbow LED prongs into the holes.

HERE'S HOW YOU STICK THE LEDS THROUGH THE MASK FROM THE OUTSIDE.

But be careful: the current must always flow through the LEDs in the right direction. Mark the point where the longer prong is (this is the positive pole) with a '+'. Push the switch into the round opening from the outside.

4. Cut one black and two red cables to a length of about 6 cm each and strip the insulation from the ends.

5. Now wire everything together as shown in the picture:

SWITCHES FROM BELOW

VIEW FROM THE INSIDE

6. Stick the battery pack (with batteries inserted) onto the inside of the box using a bit of adhesive tape.

You can switch your mask on and off with the nose switch!

This little light of mine ...

... I'm gonna let it shine

You will need:

1 battery holder
   with two 1.5 volt
   batteries
1 rainbow LED
2 crocodile cables

Plus:
1 large empty yoghurt
   pot (family size!)
1 flat polystyrene
   block
spray paint or
   acrylics
1 lantern stick
about 25 cm of string
   or wire
sharp scissors
adhesive tape
electrical tape

1. Decorate your yoghurt pot with your own design using spray paint or acrylics and cut shapes out of the sides. Do not cut the bottom 3 cm, so that the electronics can stay hidden.

2. Use sharp scissors to cut two holes on opposite edges at the top of the pot and attach a bit of string or wire to the holes to hang your lantern.

3. Spread the prongs of the LED and stick it into a polystyrene block. Mark the point where the longer prong is with a '+' for the positive pole. Connect a crocodile cable to the positive pole of the battery pack and connect it to the positive prong on the LED. Use the other crocodile cable to connect the negative pole of the battery and the shorter prong of the LED.

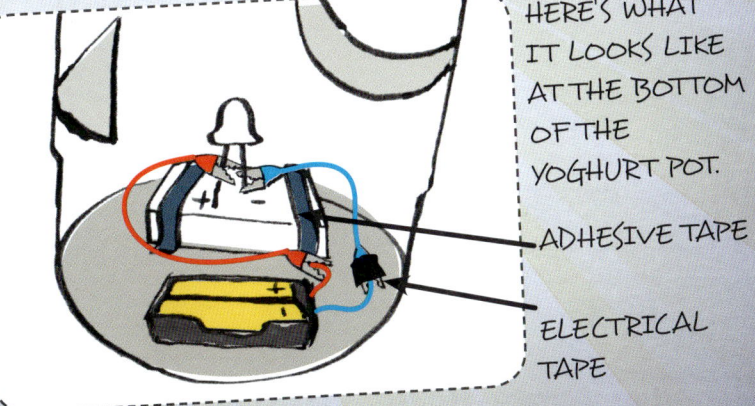

HERE'S WHAT IT LOOKS LIKE AT THE BOTTOM OF THE YOGHURT POT.

ADHESIVE TAPE

ELECTRICAL TAPE

4. To prevent a short circuit, make sure that the contacts do not touch each other. Carefully wrap a bit of electrical tape around the black connected crocodile cable and use the red crocodile cable to switch the lantern on and off.

Place the electronics carefully in the lantern and use adhesive tape to stick the polystyrene block to the bottom.

Don't want a lantern any more?

It seems a shame to waste your beautiful lantern: Just remove the string, flip the yoghurt pot, place the lighting underneath and use your lantern as a lamp instead.

Fun & action

We want to play!

So let's make our own
toys and games …

Give me an A!

Home-made game spinner

You will need:

- 1 motor
- 1 fitting
- 1 cardboard disc
- 1 battery holder
  with two 1.5 volt
  batteries
- 2 crocodile cables
- 'Town, Country, River'
  template (see page 59)

Plus:
- 1 empty tea light
- modelling clay
- 2 round drink coasters
- glue and scissors
- coloured pencils

Thinking of letters on your own for a game
of 'Town, Country, River' is yesterday's news.
Now you can use an electric spinner! It's much fairer –
you are certainly not the only one with the
suspicion that your older brother always chooses
his favourite letters!

1. Fill the empty tea light up to two thirds with modelling clay and press the motor with fitting into the clay. The upper edge of the motor should sit a few millimetres above the edge of the tea light.

2. Copy the 'Town, country, river' template on page 59. Cut out both circles, paint them and then glue each one to a drink coaster.

3. Use nail scissors to cut the inner circle out of the alphabet disc. In the other disc, use a pencil to make a hole about 3 mm big and glue the cardboard disc to the back.

4. Carefully push the alphabet disc onto the motor.

5. Now stick the second disc onto the motor fitting. The distance between the two discs should be about 3 mm. Connect one crocodile cable to the red cable of the motor and another crocodile cable to the black cable of the motor. Connect the other ends of the crocodile cables to the red and black cables of the battery pack. The upper disc then begins to spin.

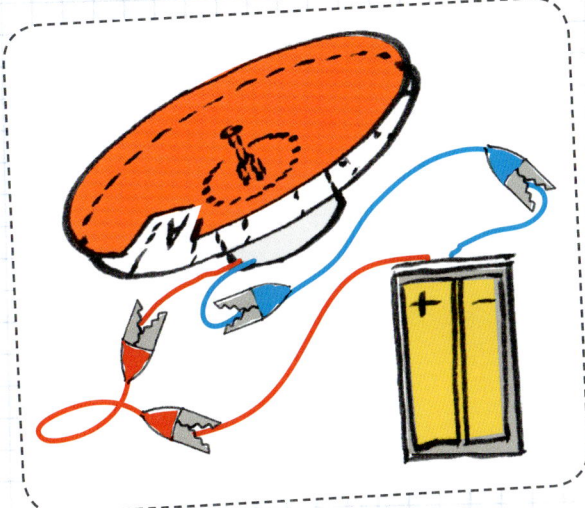

To choose a new letter for each round, connect the battery briefly so that the disc spins. Disconnect the battery and allow the disc to come to a stop. The letter at which it stops is the letter for the next round of 'Town, country, river'.

Fun tip

Make an electronic die! Instead of a disc with letters, use a disc with 6 segments and write the numbers 1 to 6 on it.

Send secret

Build a an electrical telegraph

Want to send secret messages
to your friends in another room?
Or between your two castles or forts?
Make an electrical telegraph and create
your own Morse code!

You will need:

2 bulbs
  with sockets
4 crocodile cables

Plus:
2 4.5 volt
  batteries
2 long cables with
  stripped ends
  (see page 31)
screwdriver

Create a secret code

Make a code with your friends: three short light signals
mean that enemies are on the approach.
Three long signals mean, 'You are safe to come over!'

Just like with Morse code, you can use long and short light signals
to make up your messages. Write out a list of what certain signals
mean. Use a dot to represent a short signal and a dash to represent
a long signal.

Make a mini-film ...

... and bring your dinosaur to life!

Watch T-Rex run in a mini-film,
or bring your own animations to life.
Create your own animated mini-films!

You will need:

1 motor
1 solar cell
1 cardboard disc
   with hole
2 crocodile cables
1 fitting
T-Rex template
   (see page 60)

Plus:
the bottom part of
   a cheese box, with
   a diameter of about
   10 cm
glue, scissors and
coloured pencils
double-sided
   ans normal
   adhesive tape

1. Copy the 'T-Rex' template on page 60 and cut out the marked edges. Colour in the T-Rex.

2. Create a hole about 2 mm wide in the middle of the cheese box. The hole must be exactly in the middle so that the film roll spins evenly. See page 31 to find out how to find the exact centre.

3. Use double-sided adhesive tape to stick the cardboard disc to the middle of the bottom of the cheese box.

4. Push the cheese box onto the motor axle with the fitting.

5. Use adhesive tape to stick the coloured T-Rex cut-out into the cheese box (pictures facing inwards!) so that the pictures are upright in the box.

6. Now use a crocodile cable to connect the red cable of the motor to the positive pole of the solar cell. Connect the black cable to the negative pole of the solar cell.

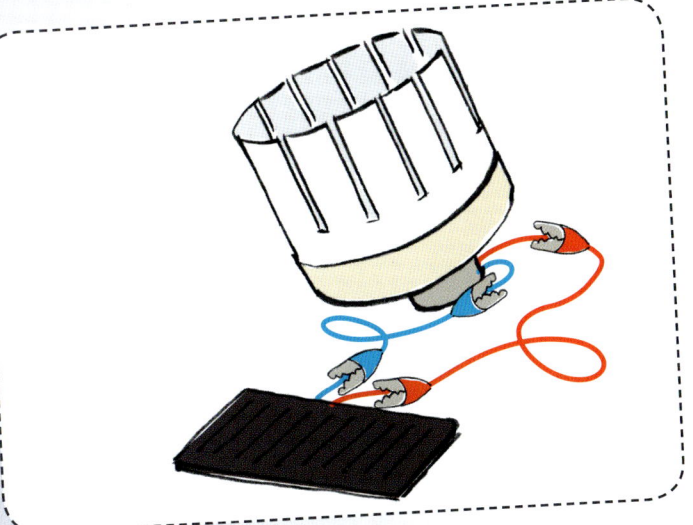

When enough light hits the solar cell, the motor will begin to move. Sometimes in artificial light, the film roll needs a little help to get moving. You can watch through the slots as the T-Rex begins to walk.

Try drawing your own animations after cutting out a new template shape. How about a running man? A film roll with a diameter of 10 cm has enough space for ten different positions for your pictures.

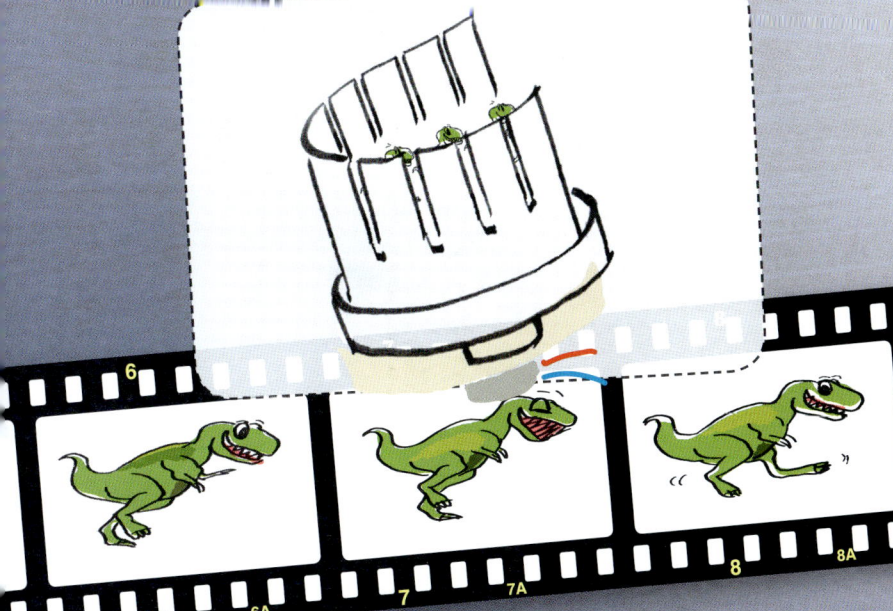

Red light!

Just don't crash ...

Use these traffic lights to regulate traffic around the construction sites and streets in your room or outside in the garden.

You will need:

1 battery holder with two 1.5 volt batteries

3 bulbs, one each of red, amber and green with sockets

2 crocodile cables

3 red and 3 black cables, about 10 cm long each

Plus:

1 empty, rectangular cardboard box

1 smaller cardboard tube

black paper or black acrylic paint

pebbles

1. Cover the cardboard box and tube with black paper. Cut three openings with a diameter of 15 mm each into the upper part of the box. On the back of the box, cut two holes with diameters of about 1 cm each opposite each big opening on the front.

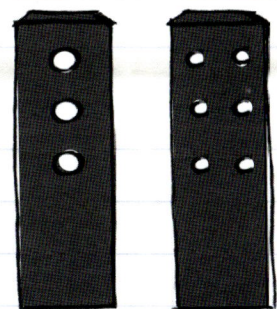

FRONT   BACK

2. Cut six bits of cable to about 10 cm in length (three red and three black) and remove 1 cm of the insulation at the ends (see page 31). Connect the stripped cable ends to the bulb sockets.

3. Now push the bulb sockets through the openings on the front of the box from the inside. Direct the cables through the openings on the back of the box from the inside. Screw a red, amber and green bulb into the three sockets.

4. Cover the cardboard tube on one end with adhesive tape and fill it to the halfway point with pebbles. Then glue the traffic light box onto the tube.

5. Place the battery pack on the back of your traffic light. To switch the red light on, connect the battery pack to the cables for the red light using crocodile cables. Do the same to the other lights to switch them on instead.

Warning!
The connection contacts for the red and black cables should not touch, otherwise a short circuit will occur.

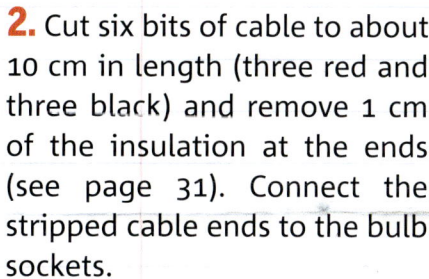

Fun tip
Your traffic lights can also be used as start signal for your next big car race!

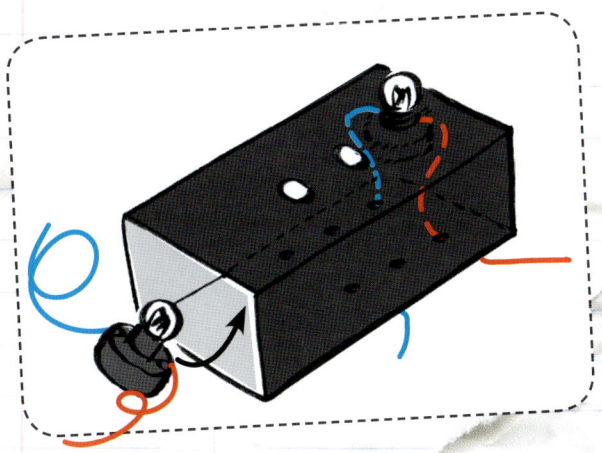

Test your friends!

Create your own quiz

Who knows the capital cities in Europe?
Who can connect the inventions to the right inventors?
The bulbs only light up when the correct answers
are chosen.

You will need:

3 crocodile cables
1 bulb with socket
1 battery holder
   with two 1.5 volt
   batteries
red cable
electronic quiz
   template (see page 61)

Plus:
paper and pencils
sturdy card, size A4
12 paper
   fasteners

5. Now you just need your test light. Use the battery pack, light bulb, socket and three crocodile cables to make the test light as shown in the picture.

1. Write out your own electronic quiz on a sheet of A4 paper. Find six pairs (such as countries and their capital cities) and write these in the wrong order in the boxes. Glue the paper to a piece of sturdy card.

2. Next to each box, stick a paper fastener through the card and bend the prongs to keep the pin in place.

3. Cut six even bits of cable and remove at least 1.5 cm of the insulation at the ends (see page 31).

4. On the back of the card, connect the answers to their correct partners using cables. Wrap the stripped cable ends around the bent prongs of the pins.

Now get a friend to connect the free crocodile cables to the correct pairs. For instance, connect a country with its correct capital city.

If your friend chooses correctly, the light will switch on. Otherwise the light will stay off, and you might have to consider giving your friend an atlas for their next birthday!

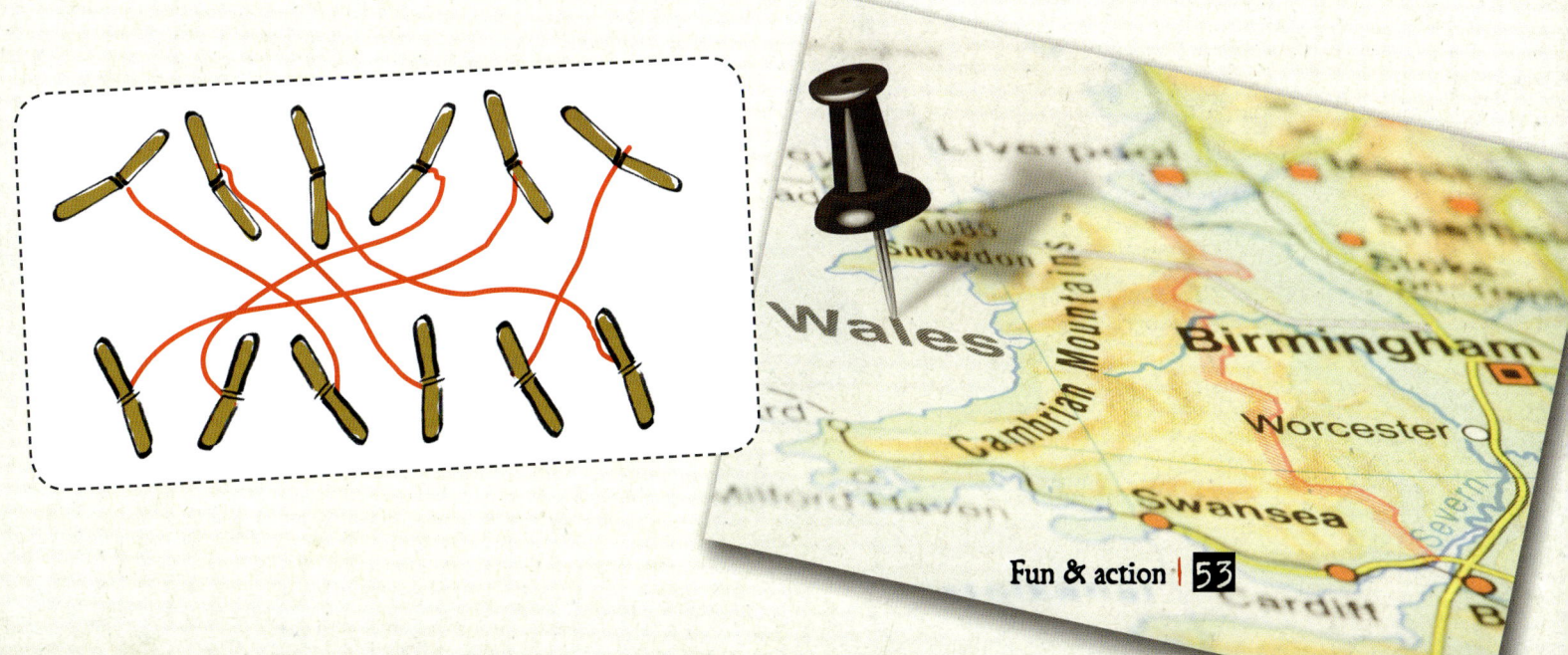

Knowledge

Electrical voltage and current

Electrical voltage runs between the positive and negative poles of a battery, and also in other power sources, such as electrical sockets on the walls.

To understand the difference between voltage and current, imagine a waterfall; the height of the waterfall is like the voltage, and the flowing water itself is like the current.

The higher the voltage between the two poles, the higher the waterfall.

Voltage is generated when lots of electrons gather at the negative pole of the battery. Electrons are negatively charged, so they just have one goal: they want to get to the positive pole as fast as possible! If the positive and negative poles are connected with a cable or wire, the electrons dash off through this cable or wire to the positive pole, creating a flow of electric current.

Closed circuit

When you connect the poles of the battery to a bulb, you create a closed circuit. The current flows through the bulb and causes it to light up.

So a circuit is really just a combination of different parts arranged in a way to achieve a specific purpose. Circuits are created for various tasks, for example they can operate motors, switch lights on and off or trigger an alarm if a burglar breaks into a secured house.

Batteries

A battery is a small, mobile power source. Each battery has a positive and negative pole. The poles are marked with the '+' or '-' symbols on the battery.

A large number of negatively charged electrons collect at the negative pole, whereas the positive pole has too few electrons.

The difference between the number of electrons at the two poles is the electrical voltage.

The higher the voltage, the more current will flow from one pole to the other.

ELECTRONS

+

If all electrons have flowed from one pole to the other, the battery is empty. Of course it is not actually empty, but now all of the negative electrons have moved from the negative pole to the positive pole, so there is no longer any battery voltage, which we need in order to operate a circuit.

Once it is empty, a battery can no longer be used if it is non-rechargeable. These batteries must be specially recycled and not thrown away with the normal household rubbish, because they contain materials that could hurt the environment.

Rechargeable batteries can be used again. When you charge your mobile phone, for example, you connect it to a plug that restores the voltage in the battery of the phone.

Conductors and non-conductors

Electrons cannot flow well through every material. Materials through which electrons can flow freely, allowing the current to flow well, are called 'conductors'.

Some examples of conductors are metals such as copper, aluminium, iron and even gold and silver.

The opposite of a conductor is a non-conductor. Electrons cannot flow well, or they sometimes cannot flow at all, through this type of material. Typical non-conductors are wood, porcelain, ceramic, glass or plastic.

These non-conductors are important for keeping the current flowing only through the intended paths. For this reason non-conductors are often used on purpose as insulators. Insulators ensure that the current only flows through the conductor and not through anything else.

Switches

Switches are very important for circuits, because almost every circuit needs to be disconnected at some point. The function of a switch can be compared with a drawbridge; when the drawbridge is down, you can cross it without problems. If the drawbridge is up, you cannot cross.

ON/OFF SWITCH

Electronic parts

Electronic parts or components are essential for an electrical circuit or assembly.

Each component has a specific task to complete in a circuit. Bulbs, for example, have quite an obvious function: they should light up. Other parts, such as transformers, resistors or capacitors, have more complicated functions, but each one has a specific task in a circuit.

Cable colours

Cables have different colours, and not just to keep electricians from getting bored! The cable colours have specific meanings:

red stands for positive, **blue** or **black** for negative.

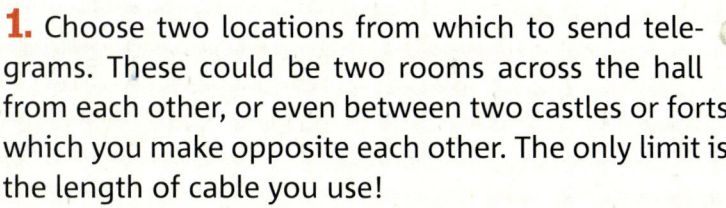

DOT - DASH - DASH

1. Choose two locations from which to send telegrams. These could be two rooms across the hall from each other, or even between two castles or forts which you make opposite each other. The only limit is the length of cable you use!

2. Create an electrical circuit at each location by connecting a bulb to a battery as shown in the picture.

3. Now connect both circuits using two long cables between the bulbs. If your cables are not long enough, you can extend them by cutting more lengths of cable, removing the insulation from the ends, twisting the ends of individual cables together and then wrapping the connection points with electrical tape.

Note:

Use two brand-new 4.5 volt batteries for best results, so that both circuits have the same charge. Also make sure that you don't send messages at the same time. Agree on a 'stop' signal to indicate when one person has finished their message, like when other devices such as walkie-talkies are used.

Knowledge

Short circuit

A short circuit occurs when the poles of a battery, for example, are directly connected to a wire without using a bulb or motor as a consumer for the power generated.

Without these consumers, a high amount of current is able to flow freely, and the battery quickly becomes empty and produces a lot of heat.

A short circuit is not so bad with a battery, but can cause electrocution or fires when it occurs with higher voltages, such as at an electrical outlet on the wall. For this reason, mains outlets in houses or cars have a fuse that blows when the current gets too high, which disconnects the circuit.

This sounds complicated, but it really isn't. Imagine two mill wheels in a stream. If the wheels are positioned one behind the other, in a row or series, the same amount of water flows through both wheels. But the water flows through the second wheel with less force than through the first wheel. The second wheel turns more slowly than the first.

If the two mill wheels are placed next to each other, in parallel, the water flows through both wheels with the same amount of force, and both wheels turn at the same speed.

In electronic circuits, the two variants also have different results. The same amount of current flows through two bulbs connected in series (one after the other), but the bulbs will not light up very brightly because the bulbs have to share the battery voltage.

WARNING!
DO NOT CONNECT THE
CABLES LIKE THIS!

SERIES CIRCUIT

However, if the two bulbs are connected in parallel to the battery, they no longer have to share the voltage and both bulbs light up with full brightness.

Parallel and series connection

Can one power source supply more than one consumer? Yes! In some projects in this book, such as the robot mask or spaceship, one battery pack supplies power to two bulbs or LEDs.

There are two ways to supply power to more than one consumer in one circuit: parallel and series connection.

PARALLEL CIRCUIT

VISITORS WELCOME

KEEP OUT!!!

Solar-powered boat pirate flag, page 20

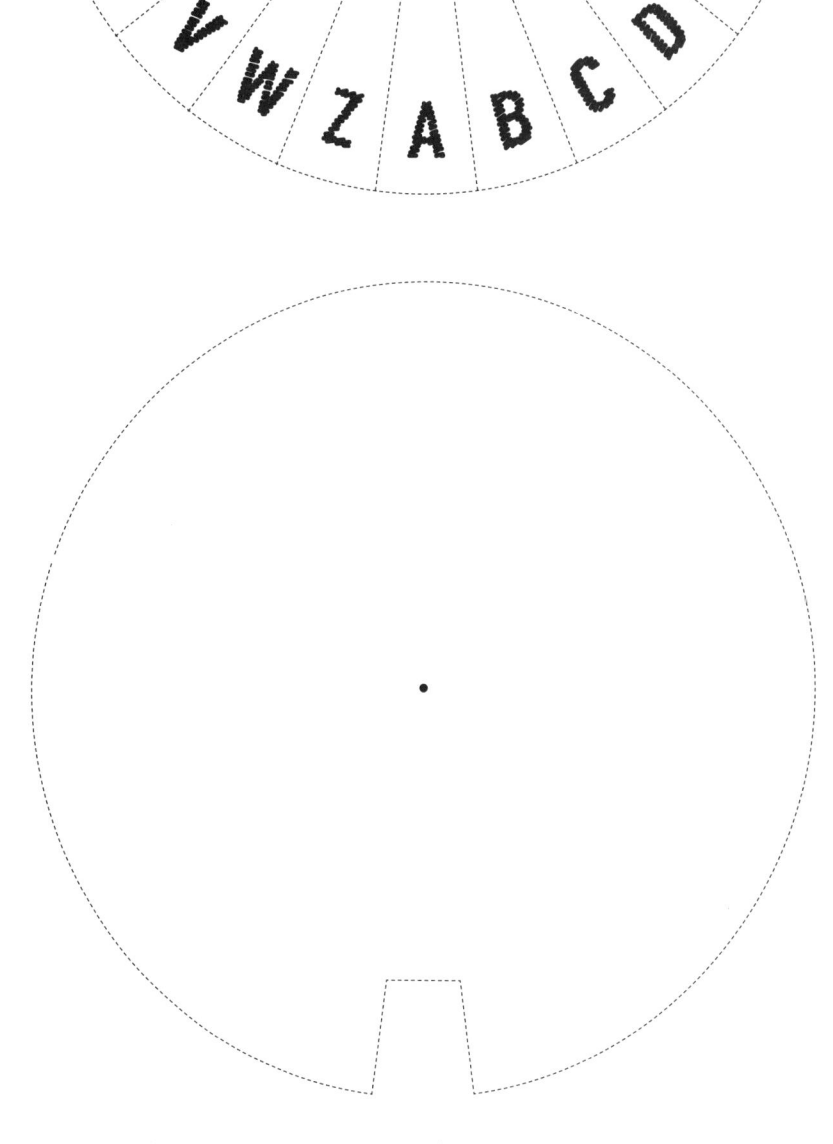

'Town, Country, River' alphabet disc, page 44

Dinosaur or runner for mini-film, page 48

Adjust the scale when copying these templates to suit the size of your cheese box and glue together.
Get an adult to help you.

Electronic quiz, page 52

Your Electronic Adventures have not come to an end yet!

We bet you have lots of better ideas for building things using the parts in this book. An even cooler electric car. An even spookier mask. An even faster boat. Show others what you've made by uploading a photo of your project onto our website: **www.abenteuerelektronik.de**. There are lots of other models to see there. And who knows, your idea might find its way into the next book in the Electronic Adventures series!

© 2013 Franzis Verlag GmbH, 85540 Haar bei München · www.elo-web.de

Author: Carmen Skupin
Idea/design: Carmen Skupin, Michael Büge, Thomas Käsbohrer, Martin Koschewa
Copy editor: Eva Wöbb
Translation provided by Surrey Translation Bureau, Farnham, Surrey, United Kingdom
Art & design cover: www.ideehochzwei.de
Content: www.evaschindler.de
Photos: www.sarahbrueck.com
The Editorial Team would like to thank the following models: Alicia, Constantin, Lea, Luca, Manuel, Simon, Viktoria and Yannik.

Illustration credits
www.fotolia.de: pg. 2 and pg. 26: Alx, pg. 8: yvart, pg. 9: HandmadePictures, pg. 10/11: Nicemonkey, Igor Kovalchuk pg. 20: ronstik, pg. 21: vector-graphic82, pg. 22: hs-creator, pg. 23: fffranz, pg. 26, pg. 28 and pg. 40: THesIMPLIFY, pg. 28: picsfive, pg. 29: by-studio, pg. 36: 007, GHotz, electriceye, petrafler, iadams, pg. 44: Jeanette Dietl, pg.48: DM7, pg. 49: typomaniac, pg. 50: Kramografie, attltibi, pg.52: Tom-Hanisch, pg. 55: Helmut Niklas, Klecks: Kirsty Pargeter. DOVER: pg. 11 and pg. 22/23. Alfred Karpf: pg. 16/17 and pg. 32.

ISBN 978-3-645-65200-1

This product was made in accordance with applicable European Directives and thus bears the CE symbol. A description of its proper use can be found in the accompanying instructions. In the case of any other use of or change to this product, you alone are responsible for following applicable rules and guidelines. Follow the supplied instructions very carefully when building your projects. This product may only be passed on to other users along with its instructions and safety information.

The symbol of the crossed-out rubbish bin indicates that this product should be disposed of with recycling as electrical waste, separately from domestic waste. Your local authorities can direct you to your nearest recycling point.